I0820319

The Frost is one of the first films fully created by AI.

THE FROST

WAYMARK CREATIVE LABS PRESENTS A JOSH RUBIN FILM 'THE FROST'
PROMPTING JEFF SYNTHESIZED TOMMY HERRMANN LEXI DIETZ AND ZACH POLEY ANIMATION BY MATT SESSIONS
EDITED BY ROBERT MCFALLS RODUCTION DESIGN STEPHEN LOGGINS PARKER

CREATED BY JOSH RUBIN AND JEFF SYNTHESIZED

AI IN THE WORLD

12 USES FOR ARTIFICIAL INTELLIGENCE IN ENTERTAINMENT

BLACK RABBIT BOOKS

Table of Contents

AI Is Taking Over the *Director's Chair*

1

Sometimes, people have ideas for movies. But making a movie can cost a lot of money. OpenAI's DALL-E 2 is an artificial intelligence (AI) system that creates pictures that look real. DALL-E 2 has been trained using billions of pictures and captions. Users put in descriptions and the AI makes its own images.

In 2023, the first movie made entirely with AI tools came out. It was a 12-minute short sci-fi film called *The Frost*. The story was written by a human. But AI brought the story to life.

The director used DALL-E 2. It helped come up with shots for the movie. Then another AI tool called D-ID added motion to still pictures. Details like blinking or moving lips came to life. Then the short clips were combined to make a full movie.

Was the movie flawless? Not really. DALL-E 2 had its own style. It leaned into a fantasy feel. But the style stayed the same from the start to finish. So the moviemakers went with it. The movie is sometimes creepy. The characters and details are not always "right." *The Frost* is about people struggling to survive. DALL-E 2's art added to the horror.

UNCANNY VALLEY

In 1970, a Japanese inventor came up with the term "uncanny valley." It describes the feeling of unease some people have when seeing something that is almost—but not quite—human. AI-generated art is often seen as having uncanny valley qualities.

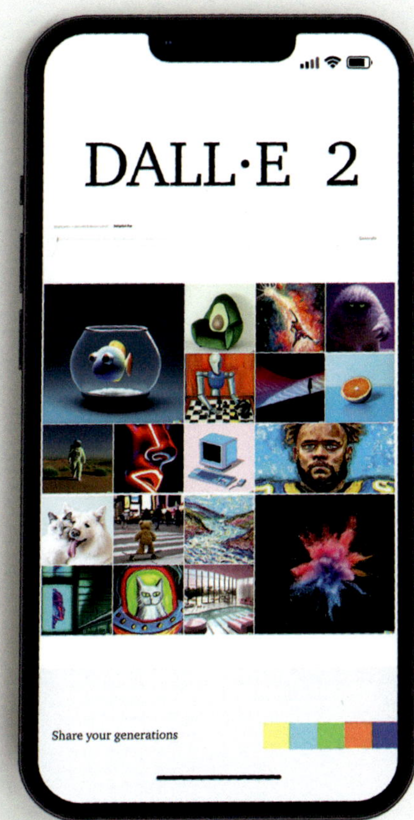

2 AI Connects Viewers to *Their TVs*

Watching TV does not require much physical or mental effort. The viewer can choose to pay close attention to the show. But they can also do other things. They might play on their phones or work. One tool, IRCODE, wants to change that. It wants people to engage with what they see.

IRCODE uses image-recognition software. It is a lot like what a **drone** uses. It also uses AI **neural network** technology. It can "see" things on a screen. Together, these create IRCODE's "EXACT Match" technology.

Users do not have to scan a QR code. They do not have to go to a new browser window. They just need IRCODE's app, LIVE+.

In 2025, LIVE+ was used during Super Bowl LIX (59). People held their phones up to the TV. Links full of information appeared. They went to things that people

AI let fans explore Kendrick Lamar's look and make purchases during the Super Bowl.

might like. Viewers could see a player's live stats. They could buy that player's jersey. They could follow them on social media.

The whole Super Bowl was scannable with the app. Viewers were sent to food menus or shopping pages. People used the app during the halftime show. They wanted to find out what Kendrick Lamar wore. IRCODE hoped viewers felt like they were part of the game.

$92 Amount an average person spent during the Super Bowl in 2025.

More than 127 million people watched the Super Bowl. • That's $18.6 billion total! • Food and beverages are the most popular purchases.

AI Helps Taylor Swift Stay *Safe on Tour*

3

Taylor Swift's Eras Tour was the highest-grossing tour of all time. More than 10 million people went to stadiums around the world. They wanted to watch Swift sing. Being with that many people can be an amazing experience. But it can also be scary, especially if you're famous.

Swift uses AI-powered **facial recognition**. She first used it during her 2018 Reputation Stadium Tour. The AI scanned faces in the crowd. It compared them to the faces of known threats. Swift has had stalkers over the years. The AI made sure they were not there.

Sparklers shoot off at a Taylor Swift concert in London in 2018.

That technology was even more advanced during the Eras Tour. For example, Accor Stadium in Australia used cutting-edge AI. It watched Swift's crowds of 80,000 fans.

AI was able to track the crowd's size. It watched where and how they moved. It made sure they were feeling happy.

A large crowd of people can quickly become unsafe. People may panic. They may start other dangerous behavior. AI is meant to keep everyone happy and calm. If people began to seem upset, announcements could be made to settle them down. More exits could be opened to let people out.

Swift and her team use AI to make concerts safer and more fun for fans.

4 Manga Goes Cyberpunk *with AI*

Manga are Japanese comic books. They can be about anything you can imagine. Some are about science fiction. There are manga about giant robots and futuristic computers. In 2023, sci-fi became reality.

Manga author Rootport wrote a story called *Cyberpunk: Peach John*. He could write, but he could not draw. He let an AI called Midjourney do that part. It was the first time a complete manga had been created by AI.

Rootport gave Midjourney descriptions of what he wanted. He had to learn what to say to the AI to change the pictures. It took him six weeks to finish. In the end, the story was more than 100 pages long. It was also in full color.

There were things the AI could not do. It could not re-create the same character. That made it hard to

1 Years Rootport estimated it would take to make *Cyberpunk: Peach John* by hand.

Rootport also included a 10-page manual. It told readers how they could make their own AI manga. • Hand-drawn manga takes a long time. Many people are involved. • It takes between 5 and 15 hours to finish a single page.

AI helps artists draw manga faster.

follow the story. So Rootport gave each character their own look. They might have pink hair or colorful clothing. They looked a little different on every page. But the reader would see pink hair. They knew which character it was supposed to be.

AI also could not make human hands look right. There are very few scenes that show hands.

KOKORO NO ME One of the first manga to use AI came out in 2018. It was called *Kokoro no Me*. The AI was made by a college professor. It could come up with story ideas. It could also make rough sketches. The author, Fujitani Youko, used the AI's ideas to create her own story.

New AI tools can turn rough sketches into polished art.

AI Turns Hollywood into *Deepfakes*

5

The first *Indiana Jones* move came out in 1981. It starred Harrison Ford. He was 38 years old. The fifth *Indiana Jones* movie came out in 2023. This time, Ford was 80.

For about 25 minutes of the movie, Ford looks like he's half that age. It was not done with makeup. They did not hire a younger actor. It was done using **special effects** and AI.

Moviemakers made a 3D model of Ford's head. They used technology called Face Swap. It took one face and put in a different one. AI watched Ford's movies. It used **machine learning**. It decided what a younger Ford's face would look like.

Then, Ford recorded his scenes. He used motion-capture cameras. Face Swap and AI added the younger faces. The tools worked together to turn Ford's performance into what movie viewers saw. It was a type of **deepfake**.

In 2024, director Robert Zemeckis also used AI. He made a movie called *Here*. He worked with an AI company called Metaphysic. Metaphysic used **generative AI**. This AI can make a digital version of an actor's body and voice.

Here took place over 60 years. Metaphysic turned Tom Hanks, Robin Wright, and the rest of the cast into younger versions of themselves. *Here* was the first time an entire movie had been built around AI visual effects.

Deepfake tech can make it look like someone said or did something they didn't.

DEEPFAKES In 2021, Metaphysic posted on TikTok. They made 10 deepfake videos. They showed actor Tom Cruise doing a variety of things. Cruise had not helped make any of the videos. Metaphysic said their videos were made for fun. But people worry that deepfakes could also be used to do harm.

2017 Year the term "deepfake" first appeared on social media.
It referred to a face-swapping code shared on Reddit. • "Deep" refers to a form of machine learning. It makes the artificial image or video. This is called "deep learning." • It is getting easier to make deepfakes as some apps have this technology.

AI Lets Viewers Write *Their Own Shows*

6

In May 2024, a company called The Simulation released 10 web series. Each one was different. There was a horror show and a comedy. There was a sci-fi show and a political cartoon.

The stories were different. But they all had one thing in common. Each had been made with Showrunner.

Users gave Showrunner prompts of 10 to 15 words. Then, it made full scenes or TV episodes. Some were only two minutes long. Others were up to 16 minutes. AI made the voices. It decided how each scene looked. It made sure the characters were the same. And it checked that the story made sense.

The creators of Showrunner hoped this would lead to two-way entertainment. That means people in Hollywood would

keep making shows. But now, people at home could too. If a show ended, fans could make new episodes. They might even put themselves in the show.

This wasn't The Simulation's first experience with AI. In 2019, the company, then known as Fable Studio, made a 40-minute **virtual reality** (VR) film. It was based on a graphic novel called *Wolves in the Wall*.

The movie's main character, Lucy, was built using ChatGPT. The VR user played the role of Lucy's friend. The film won many awards.

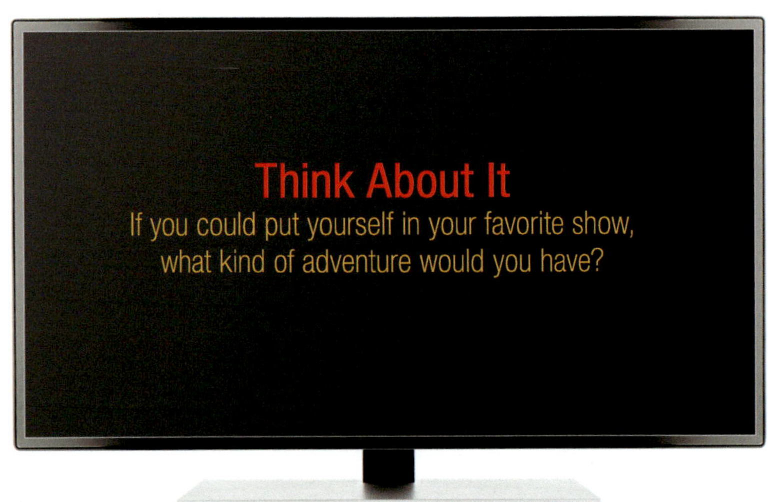

481 Number of scripted TV shows released in 2023.

That was a drop of 24 percent. There were 633 shows in 2021 and 2022. • It is expensive to make TV shows. Many shows are canceled after one or two seasons. • AI could save studios money. It could also help them create TV shows more quickly.

7

Point, Set, Match: AI Watches *the Line*

In 2002, the Hawk-Eye System was introduced. It was used during professional tennis matches. The system used computer **algorithms**. It used motion capture. It could see whether a ball was in- or out-of-bounds.

Sometimes, an umpire made a call. The player did not agree. AI would say if the call was right or not.

The COVID-19 pandemic limited how many people could be on the court at once. The US Open tennis tournament had a plan. They used Hawk-Eye cameras instead of human line judges.

The cameras create a 3D model of the court. They can tell where the ball is within a 0.1-inch (2.54-millimeter) range of error.

Hawk-Eye doesn't just know where the ball is. It shows where the ball is going. As soon as the ball hits the

8 Percent increase of accuracy of calls when Hawk-Eye was in use.

Umpires made fewer mistakes with Hawk-Eye. • Even so, almost 23 percent of balls that umpires called as "in" were actually out. • Researchers wondered if umpires felt more pressure to make the right call because of AI.

ground, Hawk-Eye calls whether it is in or out.

Tennis is not the only sport using AI on the court. The National Basketball Association (NBA) uses Hawk-Eye SkeleTRACK technology. It logs 29 points on an athlete's body. This information can be used to put players in virtual games.

In 2024, the New York Knicks and the San Antonio Spurs played each other on Christmas Day. AI made it look like the game took place in a virtual version of Disney's Magic Kingdom Park.

AI technology like SkeleTRACK can scan athletes and put them into virtual games.

AI Signs a Record *Deal*

8

In 2022, record producers thought they had found a new star. His name was FN Meka. He had green hair and face tattoos. He had more than 10 million TikTok followers. But he was an AI-powered robot rapper.

FN Meka was created by a virtual record label called Factory New. His voice came from a real artist, Kyle the Hooligan. But FN Meka's sound and lyrics were created using AI. His videos used **augmented reality (AR)**. Factory New built him using data from video games and social media.

Many popular songs are written by teams of people. They write what they think will sell. Factory New's founders think that AI could do the same thing.

In August 2022, FN Meka was signed by Capitol Records. It was the first time an AI-generated rapper had been

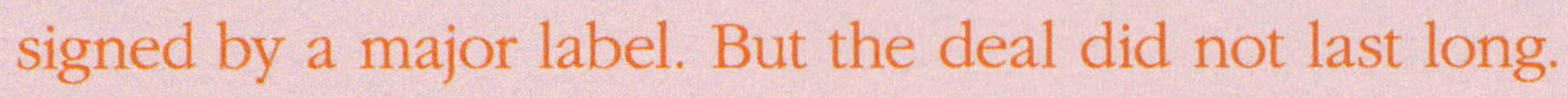
signed by a major label. But the deal did not last long.

People protested FN Meka. They said his music used harmful **stereotypes**. He was taking Black music as inspiration. But he did not live the life of a Black person. Capitol dropped FN Meka shortly after.

10 Number of days FN Meka was signed with Capitol Records.

Some people say AI spreads racist stereotypes. • AI does not understand bias. But it can pick up on it from human-created content. • Kyle the Hooligan, a real Black artist, was never paid for his voice being used in FN Meka's songs.

AI helps produce rap music by mixing beats and creating rhymes.

Think About It

How might using AI to describe someone else's lived experience be an issue?

Capital Records is based in Los Angeles, California.

AI Can Make Whatever *You Imagine*

9

The first photo on the internet showed up in 1992. Digital cameras were invented soon after. They quickly got popular. Each new version had more tools. They helped photographers take the perfect picture. And now, many digital cameras come with deep learning AI.

Some cameras have AI autofocus systems. Canon cameras tell the difference between people and objects. They can track moving people. They can focus on them even if the lighting is poor. They can find them even after the person moves behind something.

Canon's Action Priority Mode is used at sporting events. Its AI can see which players are most likely to be in the middle of the action.

The cameras learned by looking at sports photography. They also take user feedback. This helps them improve over time.

Human photographers can't be everywhere. Companies often need stock images or video. But the exact thing they want may not exist. Generative AI can create that exact thing. It just needs some help.

Adobe's Firefly program is one example. It can put two still images together to make a video. It can translate voice-overs into different languages. It can even make it sound like the same person said them.

15 billion

Number of images created using text-to-image algorithms between 2022 to 2023.

People are now creating 34 million images per day. • It took just three months for 1 billion of those images to be made with Adobe Firefly. • Around 1.8 trillion non-AI photos are taken every year.

AI can help photographers take clearer and stronger pictures.

AI Turns NPCs into *New Best Friends*

10

You are in the middle of a video game quest. But you are not quite sure what to do next.

You find a woman leading a cow. She tells you she has seen something odd. Maybe you should go to the tree at the edge of town. She may say a few other things. But she always goes back to the clue. It's helpful. It's a little predictable. Non-playable characters (NPCs) usually are.

NVIDIA is working to change that. Their **Avatar** Cloud Engine (ACE) powers their NPCs. Each character's voice, animation, and words are made by AI. The NPCs can respond in different ways. They change based on what the player does.

The technology was used in game called *Covert Protocol*. Players are detectives. They must solve a mystery by talking to NPCs. The game is different every time.

NVIDIA is a technology company known for making high-performance chips.

ACE continues to grow. Its AI is learning to watch and act like human players. NPC friends can support the player. Enemies can change the way they fight back.

NVIDIA's Audio2Face technology can be used around the world. It lets NPCs speak in more than one language. Its animation means the NPC looks natural doing it.

VR technology helps make your character look like you.

AI Lets You "Write" a *Bestseller*

11

Have you ever dreamed of writing a book? Generative AI can write it for you. Large Language Models (LLMs) are a type of machine-learning AI. Open AI's GPT-4 and Meta's Llama are two examples of LLMs. They can read through huge amounts of information. Then they can write their own.

Some people ask AI for ideas. The AI comes up with plots and characters. It thinks of story arcs. AI can also help writers organize their books. It can help with research. It can make suggestions. AI can be a valuable tool.

Other people want AI to do most, or all, of the work. There are companies that offer AI book writing services. AI looks online to learn about the subject's life. They copy the person's writing style. Then, it makes a personalized book.

There are even companies who only publish AI works. Spines was founded in 2021. It offers AI services to help write, publish, and sell books. It can finish a book in three weeks. Their goal is to help a million people become authors. TikTok and Microsoft also started similar services.

Spines says that they do not want to replace people. They just want to help everyone be an author.

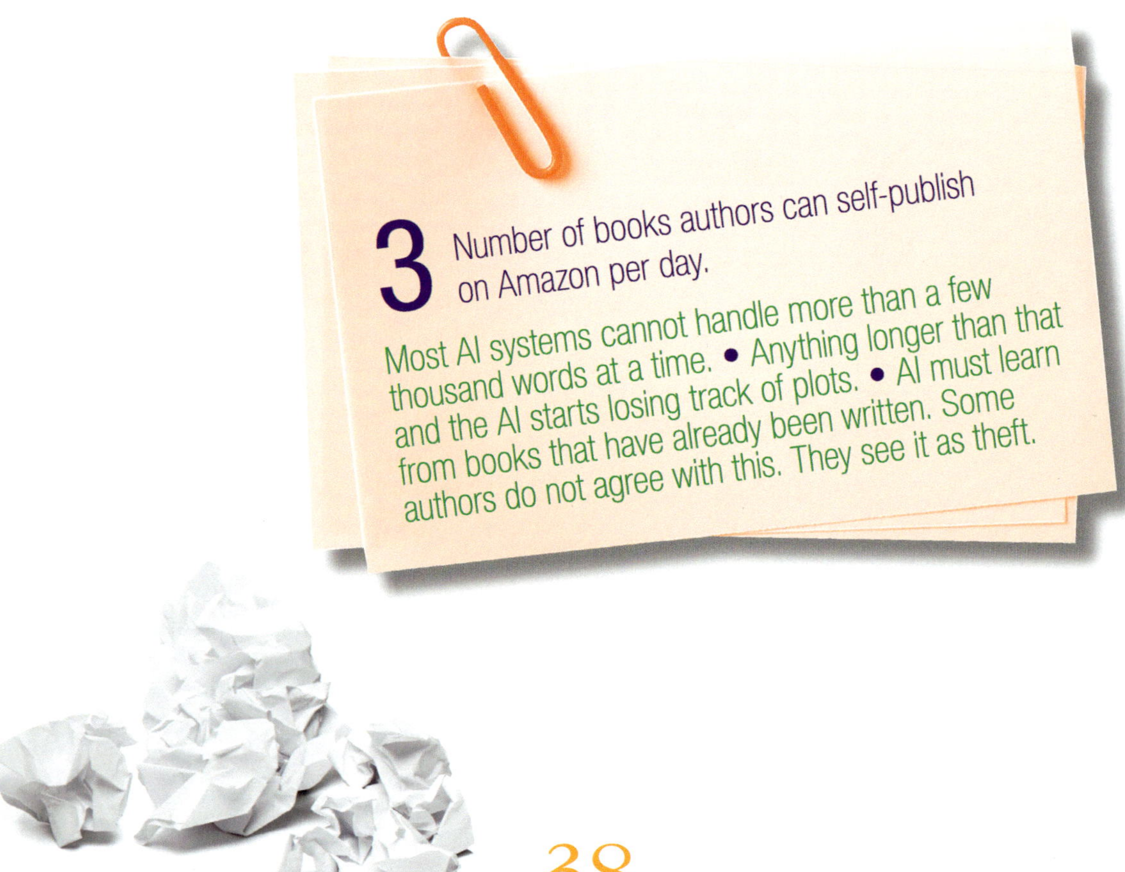

3 Number of books authors can self-publish on Amazon per day.

Most AI systems cannot handle more than a few thousand words at a time. • Anything longer than that and the AI starts losing track of plots. • AI must learn from books that have already been written. Some authors do not agree with this. They see it as theft.

Think About It

Some publishers clearly mark their books as AI. But others don't. Do you think this is a problem?

12 AI Brings Movie Effects *into Full Color*

Fast car chases. Huge explosions. Scary monsters. Exciting movie scenes keep watchers on the edge of their seat. In the past, these effects were planned and made by people. They took time. They were also expensive.

In 2024, movie company Lionsgate partnered with AI company Runway. Runway had a generative AI model. It watched more than 20,000 movies and TV shows.

With Runway's AI, directors can try new things. They can change locations without going somewhere new. They can recast an actor without having to re-shoot. They can turn human actors into cartoons.

Some people worry that using AI could harm human creativity. AI companies want their bots to be trained on **copyrighted** works. They do not want to get permission to use them. They do not want to pay, either. But learning from what has already been made is the only way AI will learn.

Actors, directors, and movie makers say that this is harmful. It steals their hard work. The actor's **union** SAG-AFTRA has fought AI. They believe actors should be asked for their permission. They should also be paid.

SEE YOU ON THE RUNWAY

Runway puts on an AI film festival. It challenges filmmakers to push the envelope using Runway tools. Creations must be under 10 minutes long. They also must include generative AI video.

AI can design daring stunts for movies without risking actors.

160,000 Number of performers who belong to SAG-AFTRA.

More than 2.3 million people work in the entertainment industry. • Actors worry that AI will create digital copies of them. Their copies will work for free. • In 2025, Vice President J.D. Vance said too many laws could kill the AI industry before it gets started.

Fact

• The NFL is using SkeleTRACK and smart watches together. Watches are synched to the game clock. Officials wear the watches. When there is a delay of game, the watches vibrate. Officials can spend more time watching what's happening on the field. They do not have to pay as much attention to the clock. SkeleTRACK can tell them if players were in or out of position. They do not need to waste time deciding whether there was a flag on the play.

• In 2024, a photo of singer Katy Perry was shared on the internet. It showed her at the MET Gala. However, Perry was not actually there. The image was created by generative AI. Even Perry's mother was tricked into thinking it was real.

Sheet

• Virtual concert avatars may change the way people experience live shows. In 2024, the band Kiss gave its final in-person concert. But the singers were turned into digital avatars. They used motion capture to record their singing and dancing. The avatars are powered by generative AI.

• In 2022, artist Jason M. Allen won an award for digital art. He used the AI tool Midjourney to create it. His win sparked an argument on the internet. On the one hand, he had made it clear that it was AI. On the other, his work had been chosen over art that had been created by people.

Glossary

algorithm
A set of steps that are followed to complete a computer process.

augmented reality
A computer technology that shows digital images on top of the current real world.

avatar
A small picture that represents an online user.

copyright
The legal right to be the only one to reproduce, publish, or sell the contents and form of a literary, musical, or artistic work.

deepfake
Videos or audio recordings that manipulate a person's likeness. They appear to do or say something they never did.

drone
A vehicle that operates without drivers inside.

facial recognition
A technology capable of matching a human face from a digital image or a video frame against a database of faces.

generative AI
A form of artificial intelligence that produces text, images, and audio.

machine learning
Enables computers to learn from data and make decisions without being explicitly programmed to do so.

neural network
A super smart computer program inspired by the human brain.

prompt
A natural language text describing the task that an AI should perform.

special effects
An image or sound used in movies or television to create images that do not exist.

stereotype
A widely believed but unfair image or idea of a particular type of a person or thing.

union
A group of workers who come together to protect their working rights and pay.

virtual reality
A digitally created experience that completely immerses the user in a simulated world.

For More Information

Books

Chambers, Ford. *AI in Entertainment.* Minneapolis: Jump!, Inc., 2025.

Harris, Chris. *Understanding Generative AI.* Buffalo, NY: PowerKids Press, 2025.

Kulz, George Anthony. *Generative Artificial Intelligence.* Parker: The Child's World, 2025.

Websites

Artificial Intelligence Facts for Kids
kids.kiddle.co/Artificial_intelligence

How YouTube Knows What You Should Watch
thecrashcourse.com/courses/how-youtube-knows-what-you-should-watch-crash-course-ai-15/

Make a Movie Recommendation System
tpt.pbslearningmedia.org/resource/lets-make-a-movie-recommendation-system-video/crash-course-artificial-intelligence/

About the Author

Mari Bolte is a writer and editor who enjoys wondering about where the future of technology will take us. Whether it's exploring outer space or thinking about how AI can make her life easier, one might say she has her head in "the cloud."

Index

TOP RANK is published by Black Rabbit Books, P.O. Box 227, Mankato, MN, 56002.

• Edited by Ana Brauer • Designed by Danny Nanos • Photographs © Alamy Stock Photo/Erickson Stock, 34, Urbanmyth, 15; Dreamstime/Amurashev, 41, Nd3000, 24–25, Zaywin Htal, 39; Getty Images/ADRIAN DENNIS, 17, Gregory Shamus, 9, Sunset Boulevard, 17, Visual Art Agency, 32; Shutterstock/ADV images, 20, Andrey Burmakin, 40, Artem Z, 30, Belinda Pretorius, 48, bombermoon, 21, Brian Friedman, 12–13, Brunohitam, 26, Carlo Prearo, 2, 28, Christian Bertrand, 11, DarioZg, 32–33, Diego Thomazini, 7, gguy, 35, Jacob Lund, 36, Jolygon, 46–47, Kundra, 2–3, 8, Luca Lorenzelli, 19, Mark Poprocki, 21, MarbellaStudio, 16, Max Acronym, 11, N Universe, 23, New Africa, 27, Paolo Bona, 24, ra2 studio, 36, Roman Samborskyi, cover, 1, Roman Sigaev, 39, Ruggiero Scardigno, 31, ruigsantos, 29, Sky Cinema, 44, Stefano Tammaro, 35, Stocksnapper, 37, 38, Suradech Prapairat, 22, Susan Montgomery, 44, Tony Norkus, 45, Tutatamafilm, 14, wjarek, 42–43, Yes058 Montree Nanta, 10, 3Dalia, 6; Wikimedia Commons/Midjourney, 4, 5, 45 • Printed in the United States of America.

Library of Congress Cataloging-in-Publication Data: Names: Bolte, Mari author | Title: 12 uses for artificial intelligence in entertainment / by Mari Bolte. | Description: Mankato, MN: Top Rank, [2026] | Series: AI in the world | Includes bibliographical references and index. | Audience: Ages 9–13 | Audience: Grades 4–6 | Identifiers: LCCN 2025021456 (print) | LCCN 2025021457 (ebook) | ISBN 9781645825173 library binding | ISBN 9781645825357 paperback | ISBN 9781645825531 ebook | Subjects: LCSH: Artificial intelligence—Juvenile literature | Artificial intelligence in motion pictures—Juvenile literature | Artificial intelligence—Music applications—Juvenile literature | Classification: LCC Q335.4 .B63 2026 (print) | LCC Q335.4 (ebook) | DDC 302.230285/63—dc23/eng/20250621 | LC record available at https://lccn.loc.gov/2025021456 | LC ebook record available at https://lccn.loc.gov/2025021457